BUILDING AI

NANCY DICKMANN

BROWN BEAR BOOKS

Published by Brown Bear Books Ltd
4877 N. Circulo Bujia, Tucson, AZ 85718
USA
and
Studio G14, Regent Studios, 1 Thane Villas,
London N7 7PH, UK

ISBN 978-1-83572-043-1 (library binding)
ISBN 978-1-83572-049-3 (paperback)
ISBN 978-1-83572-055-4 (ebook)

Library of Congress Cataloging-in-Publication Data available on request

Text: Nancy Dickmann
Consultant: Matthew Lugg
Design and Illustrations: Square and Circus
Design Manager: Keith Davis
Children's Publisher: Anne O'Daly

Picture Credits
The photographs in this book are used by permission and through the courtesy of:
Cover: Freepik.com: **Interior:** iStock: FG Trade 14–15, Mixetto 12–13; Shutterstock: Frame Stock Footage 18–19, lv-olga 20–21, James Jin 16–17, Peopleimages.com - Yuri A 8–9, Rido/Ground Picture 6–7, SeventyFour 10–11, Konstantin Yuganov 4–5. Artwork: Freepik.com.
All other artwork and photography © Brown Bear Books.

Brown Bear Books has made every attempt to contact the copyright holder. If you have any information about omissions please contact: licensing@brownbearbooks.co.uk.

Websites
The website addresses in this book were valid at the time of going to press. However, it is possible that contents or addresses may change following publication of this book. No responsibility for any such changes can be accepted by the author or the publisher. Readers should be supervised when they access the Internet.

Words in **bold** appear in the Glossary on page 23.

Manufactured in the United States of America
CPSIA compliance information: Batch#AG/5666

CONTENTS

WHAT IS AI?

Your brain controls your body. It thinks and learns. It stores memories. Scientists are trying to use computers to do the same thing. This is artificial intelligence. It's called AI for short. It's a computer **program** that can think like a human brain.

USING AI

We are already using AI. There are many different kinds. Some AIs write reports. Others make pictures or music. Some help **analyze** numbers. They read medical **scans** and search for distant planets. AI can do a lot of different jobs.

Does your TV suggest movies you will like? Those suggestions are powered by AI.

Everyday AI

We already use AI in our daily lives. Here are some of the things it does.

Smart speakers in the home listen for commands.

Online shops suggest products you will like.

Search engines help you find information.

Map **apps** show the best way to get somewhere.

Chatbots help to solve customers' problems.

Phones unlock by recognizing your face.

COMPUTER PROGRAMS

Some apps let you make video calls. They are a kind of computer program.

Computers need a program to run. This is a set of instructions. It tells the computer what to do. Each program has a job to do. Some programs edit photos. Others make spreadsheets. Apps are computer programs.

MAKING DECISIONS

An AI is also a computer program. It's just a very complicated one! An AI is given **data**. It learns from it. This helps it make decisions. It can **adapt** to new situations. Simple programs cannot do this. They cannot make their own decisions. They can only follow the instructions they were given.

Getting Stuck

A toy robot car has a simple program. If it hits something, it will change course. But what if there is a steep drop? Its program doesn't tell it what to do. It can't make its own decision to avoid it.

What's the Difference?

AI is a step up from a simple computer program. Here are some of the differences.

Simple program

- Follows instructions exactly
- Always gives the same result
- Can't adapt to new situations

AI

- Learns from absorbing data
- Gets better at its job as it learns
- Can make decisions on its own

WHAT'S THE JOB?

Your brain can do lots of things. It thinks and makes decisions. It takes in information from your eyes and ears. It keeps your body running. It controls your heart and digestion. It does this all at the same time. Brains are very good at multi-tasking!

CHOOSING A FOCUS

Computer programs and AIs are different from brains. They are designed to do one thing. The first step in designing an AI is deciding what it will do. What problem will it solve? Is this something a program can do? Could AI do it faster than a human? Could it do it better?

Your brain helps you play sports. Your eyes and feet work together.

Jobs for AI

Here are some jobs for AI.
Each program is different.
It's designed for just one job.

Looking at medical scans to find disease

Testing chemicals to find new drugs

Stopping crime by spotting suspicious activity

Helping cars drive by themselves

Making animations for computer games

Checking crops to decide what they need

TRAINING AN AI

An AI needs information. To build one, you must feed it. The information is called training data. The AI learns from it. It looks for **patterns**. That helps it understand. It's like showing the AI lots of examples. We sometimes learn like this too.

USING DATA

People gather data to feed the AI. They need a lot of it! They get rid of bad examples. They make sure everything is labeled. That way the AI knows what it's looking at. For example, every photo gets a label. It says what the photo shows.

Artists look at lots of different work. This helps them learn to make their own.

Data Everywhere

There are many kinds of data. Here are just a few of them.

TEXT DATA

Used for: Learning to understand language. This teaches an AI to write.

IMAGE DATA

Used for: Learning to recognize objects.

NUMBERS AND TABLES

Used for: Learning to see patterns in numbers.

AUDIO DATA

Used for: Learning to recognize human speech. Then we can use voice commands.

The Right Data

An AI needs a lot of data. But it must be good data. It should show a wide range of examples. It all needs to be correctly labeled. If not, the AI will not learn well.

MACHINE LEARNING

Humans learn all the time. You learn at school and at home. Parents and teachers show you how to do things. You don't always get it right the first time. But if you keep trying, you'll improve. Computers can learn this way too. It's called **machine learning**.

SELF-PROGRAMMING?

People can write programs for computers. The programs tell the computer exactly what to do. It can't do anything else. Machine learning is different. It lets the computer learn from its training data. It's like the computer learns how to program itself.

Some skills take a lot of practice. The more you work, the better you'll get.

How It Works

There are different kinds of machine learning. Here are the basic steps.

1. Gather the training data. Make sure it's labeled.

2. Write a program for the AI. It tells it how to look at the data.

3. Feed the data to the AI.

4. The AI doesn't look at the labels. But it guesses what each would be.

5. It checks the labels. What did it get wrong? It changes itself so that more guesses will be right.

6. Soon it has learned the patterns. It can guess right without labels.

ALGORITHMS

You could write an algorithm for brushing your teeth.

Have you ever baked a cake? You probably followed a recipe. It tells you what to do. It shows the steps in the right order. If you follow the recipe, the cake should be tasty. Someone else can use the same recipe. Their cake should turn out the same.

FOLLOWING STEPS

A recipe is a simple **algorithm**. Algorithms are sets of instructions. AIs use them too. That's how they learn from their training data. The algorithm tells the AI how to study it. It tells it what to do if it makes a mistake. It helps it learn from mistakes.

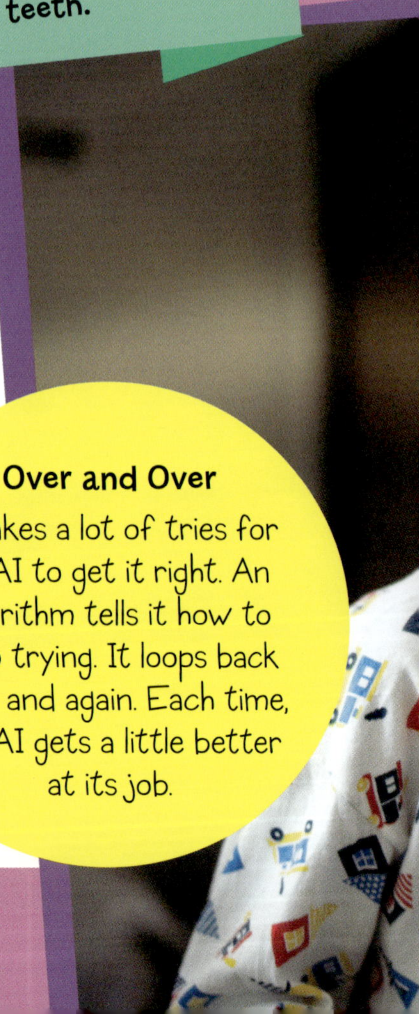

Over and Over

It takes a lot of tries for an AI to get it right. An algorithm tells it how to keep trying. It loops back again and again. Each time, the AI gets a little better at its job.

Step by Step

Here is an algorithm for walking the dog. You can put it in a flowchart. This makes it easier to follow.

COMPUTER LANGUAGES

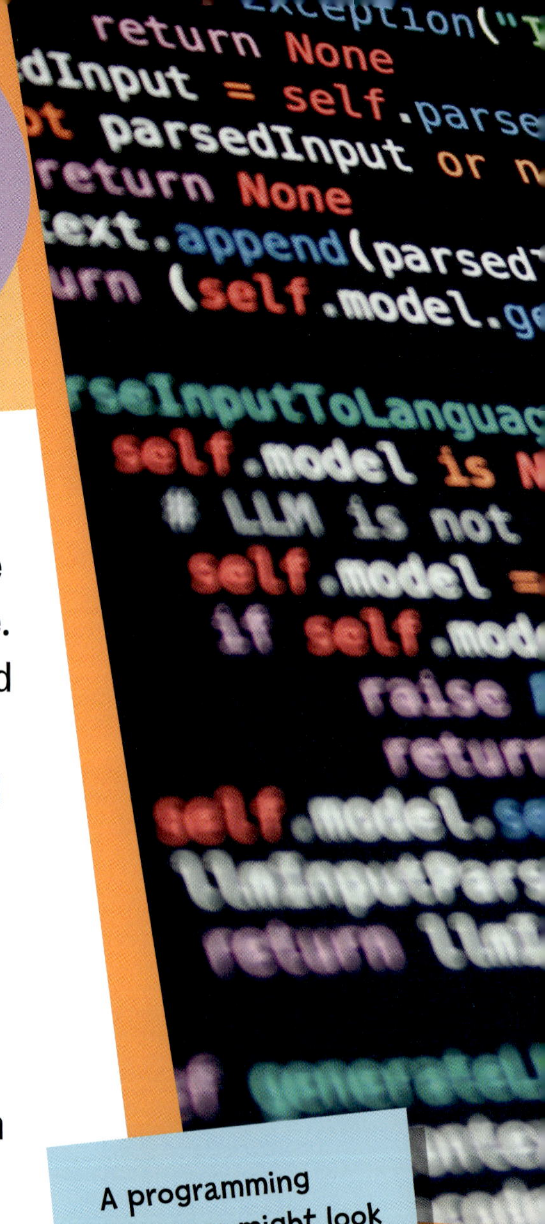

You can chat online with some AIs. You type in text. The AI replies in the same language. English is a human language. The AI has learned to use it. But it wasn't built in English. We use different languages to build AI. They are called programming languages.

COMPUTER CODE

There are many different programming languages. We use them to write algorithms. These languages use words and symbols. Each one has a meaning. Python is one of the most popular. It is often used for building AI. Java and C++ are often used too.

A programming language might look like nonsense. But the computer can understand it.

Making Sense

How does a computer understand what we want? It takes a few steps.

1. A **programmer** decides what they want the computer to do.

2. They write instructions in a language such as Python.

3. A program turns this into ones and zeroes. That is what a computer understands.

4. The computer follows the instruction in ones and zeroes.

TEST IT OUT

Your teacher sometimes gives you a test. You might not get everything right. But this shows how well you've learned. Now the teacher can show you the right answers. They know which topics to review. This will help you do better on the next test.

MAKING THE GRADE

We test AIs, too. This means giving them new data. The new data is like the training data. But it has no labels. The AI hasn't seen it before. Can it spot the patterns? Can it recognize what it sees? If not, maybe it needs better training data. The algorithm might need adjusting.

Testing shows if something works well. Cars are tested to make sure they are safe.

Troubleshooting

AI doesn't always get things right. Here are some reasons it might get things wrong.

There wasn't enough training data. This makes it harder to see patterns.

The training data wasn't balanced. There were too many of some examples, and not enough of others.

The training data didn't give enough different examples.

The algorithm didn't give clear enough instructions.

READY TO LAUNCH

Say that an AI passes all of its tests. What happens next? It is ready to launch! That means that people can now use it. There are some AIs that anyone can use. Others are for companies. They use them on their own systems. Some AIs are for researchers to use.

AI is a growing part of our lives. Self-driving cars like this may become more common.

KEEPING UP TO DATE

Once an AI is launched, there is still work to do. People must keep checking it. Is it working well? Are there problems that still need fixing? Remember that things change fast in the real world. The AI may need updates or new data.

Getting It Wrong

AI is definitely not perfect! It can make mistakes. One fast-food chain tried using AI in its drive-throughs. But the AI kept getting orders wrong. The restaurant stopped using it.

The Next AI

AI is really good at some tasks. Can you think of a problem that AI could help solve?

What kind of AI would you design?

QUIZ

How much have you learned about building AI? It's time to test your knowledge!

1. Which part of the body is AI most similar to?

a. brain

b. heart

c. stomach

2. How is AI different from simpler computer programs?

a. it's written in a programming language

b. it can make its own decisions

c. it never makes mistakes

3. What do you call the information that you feed an AI?

a. computer food

b. training wheels

c. training data

4. Which of these words means a set of instructions?

a. algorithm

b. algebra

c. alligator

The answers are on page 24.

GLOSSARY

adapt to change in order to become better suited to a new situation

algorithm a set of step-by-step instructions for performing a task

analyze to examine something in detail in order to find patterns or explain it

app a computer program designed to do a particular job, usually on a smartphone or tablet

chatbot a computer program designed to have conversations with humans

data information that is stored or used in a computer, in the form of a series of ones and zeroes

machine learning a method that lets a computer learn to do a task by analyzing lots of data

pattern an arrangement of similarities or trends between different items in a set

program a set of coded instructions for a computer to follow

programmer a person who uses computer languages to write programs for computers

programming language a language made up of words and symbols that people can use to write computer programs

scan an image taken of part of the body to help doctors spot injury or disease

FIND OUT MORE

Books

Artificial Intelligence. Julie Murray, Abdo Books, 2021.

Coding Algorithms. Samantha S. Bell, Abdo Books, 2024.

How AI Works. Lisa Idzikowski, Lerner Publications, 2025.

Websites

bbc.co.uk/bitesize/articles/zqrq7ty#zsrxhcw

kids.britannica.com/kids/article/artificial-intelligence/390648

royalsociety.org/news-resources/projects/machine-learning/videos-and-background-information/

INDEX

Answers: 1. a; 2. b; 3. c; 4. a